Jânio Kleiber Camelo de Souza
José Dantas Neto
Maria Betânia Rodrigues Silva

Fertigation of sugar cane with vinasse

Jânio Kleiber Camelo de Souza
José Dantas Neto
Maria Betânia Rodrigues Silva

Fertigation of sugar cane with vinasse

The use of vinasse on sugar cane

ScienciaScripts

Imprint

Cover image: www.ingimage.com

This book is a translation from the original published under ISBN 978-613-9-65509-0.

Publisher:
Sciencia Scripts
is a trademark of
Dodo Books Indian Ocean Ltd. and OmniScriptum S.R.L publishing group

120 High Road, East Finchley, London, N2 9ED, United Kingdom
Str. Armeneasca 28/1, office 1, Chisinau MD-2012, Republic of Moldova, Europe
Printed at: see last page
ISBN: 978-620-7-78016-7

SUMMARY

SUMMARY

The use of vinasse through fertigation is one of the most important practices, since its rational use makes it possible to fertigate arable soils, thus giving a correct destination to this product that was previously discarded in surface waters, causing pollution of these and also of the water table through percolation. As soon as the waste was properly disposed of, studies began to be carried out, making it possible to enumerate the benefits it would bring to sugar cane cultivation. The aim of this study was to survey the application of liquid waste from sugar-alcohol production (vinasse) in sugarcane cultivation. A pattern of dosage, volume, application cycle, agricultural maintenance and continuous monitoring was adopted, taking into account the climatic and geological variations that the crop covers in the region. It was concluded that the agricultural use of vinasse and its benefits for the soil are diverse, both from an agronomic, economic and social point of view. The benefit obtained through the rational use of this residue in sugarcane plantations is the increase in productivity, which occurs more intensely in poorer soils and drier regions.

Keywords: waste, sugar cane, irrigation, vinasse.

1. INTRODUCTION

The agricultural use of vinasse produced by the sugar-alcohol industry has undergone numerous changes over the years, because the continued use of vinasse in the same soils, even at low dosages, can lead to the saturation of cations, mainly potassium, causing leaching problems and subsequent contamination of groundwater. Alternatives for the rational use of this residue would be its dehydration and stabilization, however, the high energy cost of concentrating vinasse. Regardless of the process, this is perhaps its main constraint.

As such, the environment and development need to be interdependent for a harmonious sustainable process to take place, so the search for a better quality of life is the aspiration of all populations. The objective of mitigating greenhouse gases is achieved by introducing projects that result in a reduction in gas emissions or an increase in the neutralization of CH_4/CO_2, by investing in more efficient technologies or by replacing fossil energy sources with renewable ones.

Brazil is the world's largest producer and exporter of sugar cane (*Sacharum oficinarum* L.). Conab estimates that sugar production in the 2010-2011 harvest will reach 38.1 million tons, compared to 33 million tons in the previous harvest. According to the Ministry of Agriculture, 11.1 million of this total will be destined for domestic consumption. Alcohol production is expected to total 28.4 billion liters, of which 20.2 billion will be hydrous and 8.2 billion anhydrous.

Sugar cane production has been greatly encouraged in Brazil in recent years. Sugar cane production in the 2010/11 harvest is expected to total 651.5 million tons, an increase of 7.8% over the previous harvest. Compared to the first survey, the estimate is 2% **lower.** This is due to the need to replace non-renewable energy with renewable energy, promoting the continued growth of the planted area.

The final yield of sugar cane is greatly affected by the management of cultural practices on the plantation and one of the factors that most interferes with the final yield is the nutrient content of the soil and, consequently, the application doses. According to Zambello Jr. et al (1981), fertilization through irrigation plays a highly important role in increasing sugar cane productivity. The sugar-alcohol sector has made Brazil the world's largest producer of sugar cane and the only country in the world to implement an alternative fuel to petroleum on a large scale.

Today, ethanol is recognized worldwide for its environmental, social and economic advantages, and first world countries are already interested in our technology. In the 2007/08 harvest, 24 billion liters of alcohol were produced and it is estimated that the alcohol industry generated 312 billion liters of vinasse during this period (WALTER, 2008). The national sugar and alcohol industry has 350 active industries, 91 of which are independent distilleries and 259 sugar/alcohol plants, and 1,000,000 direct and indirect jobs throughout Brazil (CENBIO, 2003).

The state of São Paulo is the main national producer of sugar, alcohol and sugar-alcohol by-products. In the 2001/02 agricultural harvest, it was responsible for around 176 million tons of cane, approximately 61% of national production; 246 million bags of sugar, around 64% of national production; and 7.1 billion liters of alcohol, which represents 62% of national production (UNICA 2002).

Acidity correction and fertilization is a determining factor in crop productivity. However, this correction must be carried out in a careful and balanced manner, taking into account the use of the soil aimed at its fertility in balance with the environment.

In the Northeast, sugarcane fertilization accounts for 20% of the crop's production costs. As a result of this situation, a greater assessment of the crop's fertilizer needs is necessary (Lima Filho, 2009). Average gains of 10t cane/ha, with dosages of 300 m^3 vinasse (around a 10% increase in productivity).

Traditional agriculture in the Paraiba region is generally done without replenishing nutrients and exploited until its natural fertility is completely exhausted. This factor is aggravated by the soil and climate conditions, leading to low crop productivity. The use of vinasse will improve the physical, chemical and biological characteristics of the soil, leading to higher yields, with reduced costs and an ecologically sustainable agricultural system.

In this sense, the research aims to survey the application of liquid residues from sugar-alcohol production (vinasse) in sugarcane cultivation.

2. METHODOLOGY

The methodology of this work was based on a broad study of the literature on the subject, published scientific articles, the CETESB standard - Vinhaça - criteria and procedures for application to agricultural soil (Dec., 2006), and visits to sugar cane production units.

The methodology of this work is based mainly on the specifications of the CETESB standard (2006), which establishes the conditions of the soil to be fertiirrigated, and the parameters and formula for calculating the maximum dosage of vinasse to be applied to the soil, without causing possible pollution or sacrificing the area.

The CETESB standard (2006) recommends the following parameters to be researched for the characterization of vinasse:

- pH;

- total non-filterable residue;

- hardness;

- electrical conductivity;

- nitrate nitrogen;

- nitrite nitrogen;

- ammoniacal nitrogen;

- Kjeldhal nitrogen;

- sodium;

- calcium;

- potassium;

- magnesium;

- sulphate;

- total phosphate;

- BOD (Biochemical Oxygen Demand);

- COD (Chemical Oxygen Demand).

This characterization results in the collection of two samples at the vinasse generation site. During the harvest, the presentation of the application plan is essential for determining the K_2O content in the vinasse:

- determined the K_2O content of the vinasse, expressed in kg/m^3 .

This determination indicates the dosage of vinasse to be applied to the soil. The maximum dosage of vinasse to be applied in the treatment of agricultural soils under sugar cane cultivation was determined by the equation:

Where:

- 0.05 = 5% of the CTC;

- CEC = Cation Exchange Capacity, expressed in $cmolc/dm^3$, given by the soil fertility analysis;

- ks = potassium concentration in the soil, expressed in $cmolc/dm^3$, at a depth of 0 to 0.80 meters, given by the soil fertility analysis;

- 3744 = constant to transform the results of the fertility analysis, expressed in $cmolc/d^{m3}$ or $meq/100c^{m3}$, to kg of potassium in a volume of 1 (one) hectare by 0.80 meters deep;

m^3 of vinasse/ha = [(0.05 x CTC - ks) x 3744 + 185]/kvi Equation 1 - 185 = mass, in kg, of K_2O extracted by the crop per hectare, per cut;

- kvi = potassium concentration in vinasse, expressed in kg of $K2O/m^3$, presented in the analytical result bulletin.

To characterize the quality of the soil that will receive vinasse, a composite sample is used, consisting of four sub-samples, collected from a homogeneous plot of no more than 100 (one hundred) hectares. The sub-samples are collected, one in the center of a circle with a radius of 10 meters and the other three along the perimeter, about 120 degrees apart.

This sample is georeferenced with its coordinates: the samples are collected with an auger, continuously, to a depth of 0.80 meters. The soil is placed in a clean container specifically designed for this purpose. Once the four sub-samples have been collected, they are homogenized and a 500 gram sample is taken by quartering and sent to the laboratory for soil analysis.

The characterization of soil quality is determined on samples composed, as described above, of the following parameters:

- Al - exchangeable aluminium;
- Ca - calcium;
- Mg - magnesium;
- Na - sodium;
- SO4- sulfate;
- Dissociable hydrogen;
- K - potassium;

- Organic matter;

- CTC - cation exchange capacity;

- pH - hydrogen potential and

- V% - base saturation.

The dosage for applying vinasse to enrich agricultural soil is calculated taking into account the depth and fertility of the soil, the concentration of potassium in the vinasse and the average extraction of this element by the crop, according to Equation 1, which establishes the maximum dosage of vinasse to be applied.

The maximum concentration of potassium in the soil cannot exceed 5% of the Cation Exchange Capacity - CEC. When this limit is reached, the application of vinasse will be restricted to the replacement of this nutrient according to the average extraction by the crop, which is 185 kg of K_2O per hectare per cut.

In cases where there is a need to expand the vinasse application area, the Vinasse Application Plan must be updated. The characterization, for agricultural soil fertility purposes, of the areas that will receive vinasse application is carried out before the start of the harvest and in accordance with the procedures described for characterizing the quality of the soil that will receive vinasse application the following year.

To analyze the nitrogen (N) application rate based on the availability of this element in vinasse, the methodology recommended in CETESB standard P 4.230 of August 1999, which deals with the application of sludge from biological treatment systems in agricultural areas - criteria for design and operation, is adapted:

- the composition of the soil;
- the persistence of organic matter in the sludge;
- sludge treatment;
- the specific conditions for sludge not contaminated with pathogenic microorganisms;
- the location of the area that will receive the application;
- determining the application rate.

The lowest value calculated according to the following criteria is adopted for the application rate:

- application rate as a function of available nitrogen;
- application rate as a function of metal content;
- application rate as a function of the soil's pH-raising capacity;

- application rate as a function of other nutrients.

Soil application in tons per hectare should not exceed the ratio between the amount of nitrogen recommended for the crop in (Kg/ha) and the available nitrogen content in the sludge (Available N in Kg/ton):

To check the amount of nitrogen recommended for the crop, EMBRAPA CERRADO's recommendations for sugarcane fertilization are consulted.

3. STATE OF THE ART IN THE USE OF VINASSE

Vinasse: liquid derived from the distillation of wine, resulting from the fermentation of sugar cane juice or molasses. Known regionally as: vinasse, **vinhoto and restilo.**

3.1. Vinasse

In the manufacture of sugar and alcohol, vinasse is obtained from the fermentation of cane molasses, which is the most important by-product of alcohol distillation. The main constituent of vinasse is organic matter, basically in the form of organic acids and, to a lesser extent, cations such as K, Ca and Mg. Table 1 shows the chemical composition of vinasse.

Currently, with the increase in ethanol content in the fermented must, the average removal of vinasse (vinhoto, calda) at the base of the distillation column is in the range of 10 to 14 liters per liter of hydrated ethyl alcohol produced. Methane is produced in the anaerobic biodigestion of vinasse.

Vinasse, or vinhoto, once seen as a pollutant, now has great economic value. The National Sugarcane Improvement Program (Planalsucar), in Alagoas, has carried out studies on its chemical composition and found that organic matter expressed as free carbon is its main component; among the mineral elements, potassium stands out.

These two characteristics have made it possible to use it in the sugar cane plantation itself as an important fertilizer. Vinasse is diluted in the irrigation water of the sugar cane crop, constituting a modern irrigation process called fertigation.

Fertigation of vinasse in sugarcane plantations was intensified after bans were imposed on dumping this by-product in watercourses. In addition, this practice of applying vinasse *in natura* gained ground as it required little investment, low maintenance costs, did not involve the use of complex technology and allowed for the rapid elimination of large quantities of this material (CORTEZ, et al., 1992). Since then, some studies have indicated the beneficial action of this practice in terms of restoring some of the soil's chemical properties.

As this is a cheap and more efficient method of eliminating these residues, the dosage of vinasse applied by fertigation must be controlled.

It can be seen that the practice of disposing of vinasse on sugar cane plantations brings a viable economic return in the form of improvements in productivity, as long as the application is responsible and controlled.

Other nutrients present in vinasse are nitrogen, sulphur, calcium, magnesium and essential micronutrients for plants. It is also an important source of organic matter, which alters the soil's physical conditions, increasing the rate of infiltration and water retention, contributing to the formation of aggregates and reducing susceptibility to erosion. Over time, its constant incorporation provides considerable productivity gains.

The use of vinasse in the production of fodder yeast is widespread in Europe and other countries, where it is used as the sole source of carbon, and it is also possible to obtain potassium fertilizers from vinasse.

Table 1. Production of vinasse in Brazil in the 2001/02 harvest.

	Total cane crushed (tons)	**Total crushed cane production (50 kg bag)**	**Total alcohol production (m)[3]**	**Total vinasse produced (m)[31]**
Sâo Paulo	176.574.250	246.569.540	7.118.891	71.188.910
South Center	244.219.523	319.007.720	10.160.652	101.606.520
North East	36.822.622	64.916.980	1.400.000	14.000.000
Brazil	281.045.145	368.393.120	11.515.300	115.153.000

Source: Informaçâo ÙNICA - Year 5 - n.46 - March/April 2002.

Notes:[1] Estimated values for a proportion of 10 liters of vinasse for every liter of alcohol produced.

This waste is basically made up of all the substances introduced in the production and processing of sugar cane (water, glycerin, higher alcohols, furfural, aldehydes, succinic and acetic acid), with the exception of sugar and ethanol (Table 1). Specifically, it has a very varied chemical composition such as: nitrogen, phosphorus, potassium, oxides, sulphates, substances which are necessary for the recomposition of agricultural soils in the agro-industry (SILVA. 2006).

Table 2. Chemical composition of vinasse

DESCRIPTION	**Concentrations**			**Standard /l.alcohol**
	Minimum	**Average**	**Maximums**	
Process Data				
Wort Brix (° B)	12,00	18,65	23,65	
Wine alcohol content (° GL)	5,73	8,58	11,30	
Vinasse rate (l/l.alcohol)	5,11	10,85	16,43	10,85
Reference flow (m^3/dia)	530,00	1908,86	4128,00	
Characterization of vinasse:				
pH	3,50	4,15	4,90	
Biochemical Oxygen Demand (mg/l)	9200,00	16949,76	75330,00	175,13g
Chemical Oxygen Demand (mg/l)	10780,00	28450,00	97400,00	297,60g
Total Solids (ST) (mg/l)	260,00	25154,61	38680,00	268,90g
Total Suspended Solids (TSS) (mg/l)	40,00	3966,84	9500,00	45,71g
Fixed Suspended Solids (SSF) (mg/l)	40,00	294,38	1500,00	2,69g
Volatile Suspended Solids (VSS) (mg/l)	1509,00	3632,16	9070,00	43,02g
Total Dissolved Solids (TDS) (mg/l)	588,00	18420,06	33680,00	223,19g
Volatile Dissolved Solids (VDS) (mg/l)	921,00	6579,58	15000,00	77,98g
Fixed Dissolved Solids (FDS) (mg/l)	0,20	11872,36	24020,00	145,21g

Sedimentable Residues (RS) Ihora (ml/l)	71,00	2,29	20,00	24,81ml
Calcium (mg/l CaO)	480,00	515,25	1096,00	5,38g
Chloride (mg/l Cl)	0,50	1218,91	2300,00	12,91g
Copper (mg/l CuO)	2,00	1,20	3,00	0,01g
Iron (mg/l Fe_2O_3)	18,00	25,17	200,00	0,27g
Total Phosphorus (mg/l P_2O_4)	97,00	60,41	188,00	0,65g
Magnesium (mg/l MgO)	1,00	225,64	456,00	2,39g
Manganese (mg/l MnO)	90,00	4,82	12,00	0,05g
Nitrogen (mg/l N)	1,00	356,63	885,00	3,84g
Ammoniacal nitrogen (mg/l N)	814,00	10,94	65,00	0,12g
Total potassium (mg/l K_2O)	8,00	2034,89	3852,00	21,21g
Sodium (mg/l Na)	790,00	51,55	220,00	0,56g
Sulphate (mg/l SO_4)	5,00	1537,66	2800,00	16,17g
Sulfite (mg/l SO_4)	0,70	35,90	153,00	0,37g
Zinc (mg/l ZnO)	0,10	1,70	4,60	0,02g
Ethanol-CG (ml/l)	2,60	0,88	119,00	9.1ml
Glycerol (ml/l)	114,01	5,89	25,00	62.1ml
Yeast (dry basis) (mg/l)		403,56	1500,15	44,1g

Source: http://www.apta.sp.gov.br/cana/anexos

With these components, organic matter is dragged from the process (up to 29 COD/BOD and ranging from 20,000-35,000 mg/l), which is undesirable when disposed of in open areas, as it causes odors, and this effluent also has an acidic pH of between 5 and 5.5. The relevant federal, state and municipal environmental legislation prohibits the disposal of this effluent directly into rivers, lakes, oceans, and even randomly into the soil and air, without taking the necessary precautions.

Ideal disposal requires physico-chemical treatment and standardization of the product to ensure that it is perfectly suited to the soil's absorption capacity, thus avoiding contamination of watercourses and underground springs. The economic value that vinasse has acquired since 1975 makes this hypothesis of it being a polluting agent and consequently causing environmental imbalance questionable, given the results obtained and the data/methodology accumulated on the problem (ROSSETTO, 2004). The application of vinasse to crops as a fertilizer and in irrigation is a practice adopted by the majority of mills, with known and well-defined technology, and there are countless tests that prove the positive results obtained in agricultural productivity directly associated with the saving of industrialized fertilizers (PENATTI, 1988).

Alluvial soils are preferable for sugarcane plantations. Therefore, it is necessary to analyze and correct the soil. In this context, research shows that productivity is excellent when the plant is grown in soils with a pH of between 7 and 7.3. According to data from EMATER-PR, sugar cane grows efficiently in soils with a pH between 5.5 and 6.5, and requires correction in the case of more acidic soils.

Land preparation is one of the basic supports for good sugarcane crop yields. Another aspect to note is the intense mechanization at all stages of agro-industrial operations. This accelerates the wear and tear on the soil and the subsequent increase in inputs to adapt to the growing conditions.

The main destination of vinasse is its use as fertilizer in sugar cane plantations, and it is considered a serious environmental contaminant when used without prior treatment.

Mitigating environmental risks and adding value to biomass are the activities of this process, which is related to methane gas (CH_4) and carbon dioxide (CO_2), to the waste sector and also to the treatment of liquid effluents emitted by the sugar-alcohol industry. Discharged untreated into the ground, it contaminates water tables, affects the entire biome, destroys crops and kills aquatic life. It also encourages the rapid microbial decomposition of organic matter, which leads to the proliferation of flies, making the environment unhealthy and unbalancing the ecosystem (SMEETS, 2008).

The process of stabilizing vinasse, i.e. dehydrating it using the heat from the ethanol production process, with the recovery of evaporated water, under temperature and pressure conditions close to the environment, without the need for high investment and consequently with low operating costs, effectively eliminates methane emissions from the anaerobic fermentation of vinasse.

The contribution of using the technology mentioned in this process in agro-industries will contribute to the country's sustainable development, generating real, measurable and long-term benefits related to climate change mitigation and nutrient utilization, in a technical and economical way.

For international buyers it will be an endorsement of quality for the sugar and alcohol industries that use this process, reducing greenhouse gases critical to the ozone layer and water resources, reducing diseases caused by condensation for use in the industrial process itself or in agriculture, removing this harmful biomass from water resources (PARNADEAU, 2008).

Quantification: - Volume basically varies according to the alcohol content of the wine and the direct steam used (10 to 15 liters/liter of alcohol).

- Average volume obtained from CMAI - CTC's Mutual Agroindustrial Control:

- 1999/00 harvest 10.80 L/L	-Associated with the CTC
- 2000/01 harvest 10.32 L/L	
- 2001/02 harvest 10.20 L/L	
- 2002/03 harvest 10.41 L/L	
- 2003/04 harvest 10.46 L/L	- Associated with CTC (Canavieiro) - greater representation in the CS Region
- 2004/05 harvest 11.40 L/L	
- 2005/06 harvest 11.98 L/L	
- 2006/07 harvest 11.96 L/L	
- 2007/08 harvest 11.96 L/L	

Table 3. By taking stock of the mass in column A (the distillation column), as shown in the diagram, and assuming certain data, we can obtain an estimate of vinasse per alcohol produced.

Alcohol content of wine= 7-11^0 GL

Alcohol production 2^a = 5- 7%

Phlegm alcohol content= 50^0 GL

Alcohol content of 2^a = 88-92^0 GL

So, in terms of alcohol production, we have

7 liters/liter of alcohol (pure fermentation with a high alcohol content)

Up to 15 liters/liter of alcohol (with steam incorporated and low degree of fermentation)

Phlegm can also be incorporated into vinasse, increasing it by around 2 l/l

Elia Neto, A and Nakahodo, T. Caracterizaçâo da Vinhaça, CTC - 1995.

Table 4. Average characterization of vinasse

pH	**4**
Temperature	900
Vinasse flow	10.85L/L alcohol
BOD5	16950 mg/L
COD	28450 mg/L
Total Solids	25155 mg/L
Potassium	KG.K/ m3
Organic Load	
COD/L. alcohol	300g

Source: Elia Neto, A and Nakahodo, T Caracterizaçâo da Vinhaça, CTC - 1995.

3.2. Effect of vinasse on the soil

Glória and Orlando Filho (1983) listed the following effects of vinasse on the soil: a) raising the pH; b) increasing the availability of some ions; c) increasing the cation exchange capacity (CEC); d) increasing the water retention capacity; and e) improving the physical structure of the soil. Vinasse should also be seen as an agent for increasing microbial population and activity in the soil.

The pH of soils treated with vinasse increases (Silva and Ribeiro, 1998) mainly in areas that have been cultivated for longer, although in the first ten days after its application the pH suffers a considerable reduction and then rises abruptly, reaching values above seven; this effect is linked to the action of microorganisms (ROSSETTO, 1987).

Organic matter can be considered an important factor in agricultural productivity due to its influence on the soil's chemical, physical and biological properties (AZEREDO and MANHÂES, 1983).

When the organic matter contained in vinasse is incorporated into the soil, it is colonized by fungi, which transform it into humus, neutralizing the acidity of the medium and thus preparing the way for bacterial proliferation; Thus, when added as a fertilizer, it also favours the development of these microorganisms which act in the mineralization and immobilization of nitrogen and in its nitrification, denitrification and biological fixation, as well as microorganisms participating in the biogeochemical cycles of other elements.

According to Neves et al. (1983) the addition of vinasse, together with the incorporation of organic matter, can improve the physical conditions of the soil and promote greater mobilization of nutrients, due to the greater solubility provided by the liquid residue. Cunha et al. (1981) reported that the dynamics of nitrogen in soil treated with organic waste, such as vinasse, is complex due to biochemical transformations.

Nitrogen is present in vinasse, predominantly in organic form, and mineralization is the first biological transformation that takes place in the soil.

Since the C/N ratio in vinasse is low, significant mineralization of the immobilized forms of nitrogen in the soil can occur. However, Madejón et al. (2001), working with an association of vinasse and other organic materials in soil conditioning, observed a small increase in salinity, an increase in organic matter content and total nitrogen.

Lyra et al. (2003) agree with Madejon et al. (2001) in suggesting that an increase in the concentration of salts in the soil and a potential risk of salinization can be expected with the application of vinasse over the years, considering the rate of potassium absorption by the plant, its concentration in the vinasse and the low electrical conductivity (EC) observed in the water table, indicative of reduced leaching.

Ridesa (1994) observed that the efficiency of biochemical oxygen demand (BOD) removal by the soil can be affected by its infiltration capacity and the presence of vegetation cover and, therefore, the more vegetation cover the greater the biological removal capacity.

BOD removal efficiency has been found to be high, even in soils with high infiltration capacity. With regard to nitrate, one of the biggest problems cited in the literature is groundwater contamination.

Madejón et al. (2001) observed that, with the removal of vegetation after each harvest, nitrate levels rose in soils and concluded that its leaching below the root zone can be negligible when these compounds are applied to vegetated areas using appropriate agricultural practices, such as the use of areas far from springs or regions with a shallow water table.

Cunha et al. (1981), studying the use of vinasse as a fertilizer and soil conditioner, observed that the accumulation of potassium in the profile was not great, this element being retained in the 0.50 m depth layer, and its leaching was small, following the internal drainage in the profile.

Canellas et al. (2003), working on Cambissolo, found no increase in potassium or calcium at depth, nor in CEC or nitrogen; however, they did observe an increase in copper and iron levels in the 0.20-0.40 m depth layer. These authors also found an increase in the folic acid fraction of up to 13% in the 0.40 m layer, which could indicate both the chemical evolution of organic compounds and the transport of this fraction to deeper soil layers.

3.3. Legal use of vinasse in agriculture

In recent years, man has used natural resources with great voracity and, as a result, there has been an abundance of food, fuel and technology capable of saving lives. On the other hand, there has been intense environmental degradation, with a reduction in biodiversity, contamination of water sources and soil, among other things (SCHELP, 2005).

When the population grows as a result of society's success, the pressure for food becomes too much for natural resources. The result is famine, which leads to social breakdown and civil war. With this in mind, decrees, laws and regulations have been drawn up to protect natural resources.

According to Abreu Jùnior et al. (2005), waste from human activities has been used in agriculture for many years; however, with its increasingly intense use, public concern about environmental safety has also grown, so that legislation has been drawn up to regulate the use of this waste in agriculture.

Although Brazil does not yet have a specific national policy for the use of waste (Abreu Jùnior et al., 2005) there are already some laws and decrees that must be obeyed in order to use agro-industrial waste in the region's agriculture; for example, the Water Code (Decree 24.643 of July 10, 1934) which, among other things, protects bodies of water against the disposal of pollutants; the Forestry Code (Law 4.771 of September 15, 1965) which, among other provisions, sets a minimum limit of 20% tree cover in the southern part of the Midwest region.

According to Article 2, the minimum widths of the riparian forest strips, according to the width of the watercourses, thus not allowing large plantations to be located on their banks and in the springs, in order to minimize problems of water contamination.

There is also the environmental crimes law (Law 9.605, of February 12, 1998) which imposes the criminal penalties to be applied as a result of damage to the environment.

In the specific case of effluents, articles 33 and 54 of this law cover the most important points, clarifying the penalties that will be imposed on offenders when animals die as a result of effluents being emitted into the water and when water pollution occurs, respectively, in which case the offenders are sentenced to imprisonment of one to three years or a fine, or both, cumulatively, for violations of Article 33 and one to five years imprisonment for violations of Article 54. 54.

In the specific case of vinasse, there is also Ministry of the Interior Order No. 323 of 29/11/1978, which states that, as of the 1979/1980 harvest, the direct or indirect discharge of vinasse into any water collection by alcohol distilleries installed or to be installed in the country is prohibited.

Ministry of the Interior Ordinance No. 158 of November 3, 1980, which provides for its discharge into water collections and effluents from sugar mill distilleries; CNRH Resolution No. 15 of June 1, 2001, which sets out guidelines for the integrated management of surface, underground and meteoric waters; Ministry of Health Ordinance No. 518/04 of March 25, 2004, which establishes procedures and responsibilities for the control and monitoring of water quality for human consumption and its potability standard, as well as other provisions. 518/04 of 25/03/2004, which establishes procedures and responsibilities relating to the control and monitoring of the quality of water for human consumption and its standard of potability, as well as other measures.

There is also Law no. 9.605, of 12/02/98, which provides for criminal and administrative sanctions arising from conduct and activities harmful to the environment; Law no. 7.960, of 21/12/89, which provides for temporary imprisonment for the crime of poisoning drinking water, among others; Decree-Law no. 1.413, of 14/08/75, which provides for the control of environmental pollution caused by industrial activities; Ministry of the Interior Ordinance no. 124, of 20/08/80, which lays down rules for the prevention of water pollution, for the location of potentially polluting industries, buildings or structures and for protection devices.

And finally, the most recent CONAMA Resolution No. 357, of March 17, 2005, which provides for the classification of bodies of water and environmental guidelines for their classification, and also establishes the conditions and standards for the discharge of effluents and other measures. Paragraph 4 establishes the conditions for discharging effluents and paragraph 5 regulates the standards for discharging effluents, which are:

I - pH between 5 and 9

II - temperature: less than 40 °C, the temperature variation of the receiving body should not exceed 3 °C in the mixing zone

III - sedimentable materials: up to 1 mL L^{-1} in a 1-hour Imhoff cone test. For discharge into lakes and ponds, where the circulation speed is practically zero, sedimentable materials must be virtually absent.

IV - discharge regime with a maximum flow of up to 1/2 the average flow of the polluter's daily activity period, except in cases permitted by the competent authority

V - oils and greases

1 - mineral oils: up to 20 mg L^{-1}

2 - vegetable oils and animal fats: up to 50 mg L^{-1}

VI - no floating materials.

3.4 Forms of vinasse application

The systems currently used to fertigate sugarcane plantations with liquids, vinasse and existing wastewater are:

- **Conventional tanker and sprinkler application:** in sprinkler application, direct-mounted and self-propelled systems with a reel are used, which can be fed directly from canals or from trucks.

In the recent past, the tanker fertigation system was the most widespread way of distributing vinasse. Its advantage is the short time required to implement it, i.e. it is practically enough to acquire the fleet and put it into operation.

This may have been a determining factor in its rapid spread in the country's mills and distilleries. The existing limitations of this system include the aggravation of **soil compaction** problems**, the** impossibility of applying it in areas with plant cane, difficulties on rainy days, low distribution uniformity and the short distance over which fertigation is economically viable compared to mineral fertilization.

Vinasse, diluted with wastewater or not, can be applied by means of sprinkler irrigation systems, with the direct-mounted system and the self-propelled system with a **reel** being the ones used in Brazilian mills and distilleries.

- **Direct-mounted system***:* basically consists of a motor pump coupled to a cannon-type sprinkler, both mounted on wheeled chassis. The system can also be fitted with extensions (pipes) to increase the spacing between channels, i.e. to reduce the number of channels running through the cane fields.

The main advantage of the direct-mount sprinkler system, when compared to the tanker truck system, is the lower **cost** per unit of area fertigated. On the other hand, the main **limitation** is the need for a network of feeder channels, which implies partial systematization of the land and cuttings in the sugarcane field, when the system is not implemented at the time of sugarcane reform.

- **Self-propelled system with reel***:* this is the most widespread **system** currently used in mills and distilleries, and was introduced to replace the extension of the direct sprinkler **system.**

The main advantage of the **system** is that it is **semi-mechanized** and therefore requires less **manpower** than **direct** assembly (transporting and handling the extensions). On the other hand, the self-propelled system requires more power from the pump and therefore consumes more fuel.

Table 5. Average percentage of vinasse applied to sugar cane crops in the state of São Paulo.

Form of application	Share (%)
Conventional Tank Truck	6
Sprinkling (channel + direct mounting)	10
Sprinkling (channel + roller)	53
Sprinkling (truck + roller)	31

Table 5 summarizes the average percentage of each vinasse application system in sugarcane plantations in the state of São Paulo.

The agricultural use of vinasse produced by the sugar-alcohol industry has undergone numerous changes over the years, because the continued use of vinasse in the same soils, even at low dosages, can lead to saturation of cations, mainly potassium, causing leaching problems and subsequent contamination of groundwater. Alternatives for the rational use of this residue would be its dehydration and stabilization. However, the high energy cost of concentrating vinasse, regardless of the process, is perhaps its main constraint.

As such, the environment and development need to be interdependent for a harmonious sustainable process to take place, so the search for a better quality of life is the aspiration of all populations. The objective of mitigating greenhouse gases is achieved by introducing projects that result in a reduction in gas emissions or an increase in the neutralization of $CH4/CO2$, by investing in more efficient technologies or by replacing fossil energy sources with renewable ones.

4. FINAL CONSIDERATIONS

It was observed that the agricultural use of vinasse and its benefits for the soil are indisputable, both from an agronomic, economic and social point of view. The immediate benefit of the rational use of this residue in sugarcane plantations is the increase in productivity, which occurs more intensely in poorer soils and drier regions.

This includes fertilizer savings and society's growing concern about the risk that this highly polluting material can cause to surface watercourses (rivers, lakes, springs and wetlands) and the water table through percolation into the groundwater.

This concern is all the more latent since this type of pollution is not immediately noticed and often, when it is, there is little chance of reversing it. Many experts point to technological solutions as the main alternative for reducing the organic load of this waste and reusing it in fertigation with less risk of damage to nature.

It is also up to the cane supplier to contact the mill or distillery that supplies his production to find out the financial viability of the vinasse application technique in his cane field.

As already mentioned, there are many benefits, which can be summarized as increased productivity, longevity of the cane field, improved sprouting of ratoons, supply of nutrients and savings in chemical fertilization of the crop, among others.

5. BIBLIOGRAPHICAL REFERENCES

ABREU JÙNIOR, C. H.; BOARETTO, A. E.; MURAOKA, T.; KIEHL, J. C. Agricultural use of potentially polluting organic residues: soil chemical properties and plant production. Tópicos Especiais em Ciência do Solo, Viçosa, v.4, p.391-470, 2005.

AZEVEDO, D. F.; MANHÂES, M. S. Organic fertilization. In: Orlando Filho, J. (ed.). Sugarcane nutrition and fertilization in Brazil. Piracicaba: IAA/Planalsucar, 1983, p.227-264.

BROUYÈRE, S.; DASSARGUES, A.; HALLET, V. Migration of contaminants through the unsaturated zone overlying the Hesbaye chalky aquifer in Belgium: A field investigation. Journal of Contaminant Hydrology, v.72, n.1-4, p.135-164, 2004.

CANELLAS, L. P.; VELLOSO, A. C. X.; MARCIANO, C. R.; RAMALHO, J. F. G. P.; ROUMJANEK, V. M.; REZENDE, C. E.; SANTOS, G. A. Chemical properties of a Cambissolo cultivated with sugarcane, with preservation of straw and long-term addition of vinasse. Revista Brasileira de Ciência do Solo, Viçosa, v.27, n. 5, p.935-944, 2003.

CONAB - National Supply Company. Available at: <http://www.conab.gov.br/conabweb/index.php?PAG=133>. Accessed on: June 6, 2009.

CUNHA, R. C. A.; COSTA, A. C. S.; MASET FILHO, B.; CASARINI, D. C. P. Effects of irrigation with vinasse and dynamics of its constituents in the soil: I - physical and chemical aspects. Water Science Technology, v.19, n.8, p.155-165, 1981.

ELIA NETO, A and NAKAHODO, T. Characterization of Vinhaça, CTC - 1995.

FREIRE, W. J.; CORTEZ, L. A. B. **Sugarcane vinasse**. Guaiba: Agropecuâria, 2000.

GLÓRIA, N. A.; ORLANDO FILHO, J. Aplicação de vinhaça como fertilizante. Sâo Paulo: Coopersucar, 1983. 38p.

LIMA FILHO, O. F. History and use of sodium silicate in agriculture. Dourados: Embrapa Agropecuâria Oeste, 2009. 112 p.

LIMA, W. P. Impacto ambiental do eucalipto. 2.ed. Sâo Paulo: EDUSP, 1996. 301p.

LYRA, M. R. C. C.; ROLIM, M. M.; SILVA, J. A. A. Top sequence of soils fertirrigated with vinasse: contribution to groundwater quality. Revista Brasileira de Engenharia Agricola e Ambiental, v.7, n.3, p.525-532, 2003.

MADEJÓN, E.; LOPEZ, R.; MURILLO, J. M.; CABRERA, F. Agricultural use of three (sugar-beet) vinasse composts: effect on crops and chemical properties of a Cambisol soil in the Guadalquivir river valley (SW Spain). Agriculture, Ecosystems and Environment, v. 84, n.1, p.55-65, 2001.

NEVES, M. C. P.; LIMA, I. T.; DOBEREINER, J. Efeito da vinhaça sobre a microflora do solo. Revista Brasileira de Ciência do Solo, Campinas, v.7, n.2, p.131-136, 1983.

PARNAUDEAU, V.; CONDOM. N.; OLIVER, P.; CAZEVIEILLE, P.; RECOUS, S. Vinasse organic matter quality and mineralization potential, as influenced by raw material, fermentation and concentration processes. Bioresource Technology, n. 99, p. 15531562, 2008.

PENATTI, C. P.; CAMBRIA, S.; BONI, P.S.; ARRUDA, F.C.; MANOEL LA. Effects of applying vinasse and nitrogen to sugarcane ratoons. Copersucar Technical Bulletin, n. 44, p. 32-38, 1988.

RIDESA, R. I. P. O. D. S. Aspectos ambientais da aplicação de vinhaça no solo. Sâo Paulo: UFSCar, 1994. 67p.

ROSSETTO, A. J. Agronomic use of by-products and residues from the sugar and alcohol industry. In: Paranhos, S.B. (ed.). Sugar cane: cultivation and utilization. Campinas:Fundaçâo Cargill, 1987, v.2, p.435-504.

ROSSETTO. R., SPIRONELLO, A.; CANTARELLA. H.; QUAGGIO, J. A. Liming for sugar cane and its interaction with potassium fertilization. Bragantia, n. 63. p. 105-119. 2004.

SCHELP, D. The blindness of civilizations. Veja, v.38, n.36, p.102108, 2005.

SCHVARTZ, C. Vinasse Treatment: Anaerobic Biodigestion. In: Projeto Programa de Pesquisa em políticas pùblicas WOKSHOP TECNOLÓGICO SOBRE VINHAÇA Panel 5 IPT/CETAE/LRAC Centro de Tecnologias ambientais e Energéticas Laboratório de Residuos e Áreas contaminadas. Available at:

HTTP:/WWW.apta.sp.gov.br/cana/anexos/.

Accessed on March 20, 2009.

SILVA, A. J. N.; RIBEIRO, M. R. Characterization of a Yellow Latosol under continuous sugarcane cultivation in the State of Alagoas: chemical properties. Revista Brasileira de Ciência do Solo, v.22, n.2, p.291-299, 1998.

SMEETS, E.; JUNGINGER, M.; FAAU, A.; WALTER. A.; DOLZAN, P.; TURKENBURG, W. The sustainability of Brazilian ethanol - An assessment of the possibilities of certified production. Biomass and Bioenergy. n. 32, p. 781-813, 2008.

SZMRECSANYI, T.(1994). Technology and environmental degradation: the case of the sugarcane agroindustry in the state of São Paulo. *Revista* Informaçôes Econômicas, Sâo Paulo, Vol. 24, N°.10, October 1994.

TALSMA, T.; MANSELL, R. S.; HALLAM, P. M. Potassium and chloride movement in a forest soil under simulated rainfall. Australian Journal of Soil Research, v.18, n.3, p.333-342, 1980.

ÙNICA (2002). ÙNICA Information. Year 5, N° .46, March/April 2002.

MIX
Papier aus verantwortungsvollen Quellen
Paper from responsible sources
FSC® C105338

Printed by Books on Demand GmbH, Norderstedt / Germany